Copyright 2016 by Jamy Jackson- A

his document is geared towards provid
formation in regards to the topic an
ublication is sold with the idea that
equired to render accounting, offi
therwise, qualified services. If advice is necessary, legal or
rofessional, a practiced individual in the profession should be
rdered.

From a Declaration of Principles which was accepted and
pproved equally by a Committee of the American Bar
ssociation and a Committee of Publishers and Associations.

n no way is it legal to reproduce, duplicate, or transmit any
art of this document in either electronic means or in printed
rmat. Recording of this publication is strictly prohibited and
ny storage of this document is not allowed unless with
ritten permission from the publisher. All rights reserved.

he information provided herein is stated to be truthful and
onsistent, in that any liability, in terms of inattention or
therwise, by any usage or abuse of any policies, processes, or
irections contained within is the solitary and utter
esponsibility of the recipient reader. Under no circumstances
ill any legal responsibility or blame be held against the
ublisher for any reparation, damages, or monetary loss due to
he information herein, either directly or indirectly.

espective authors own all copyrights not held by the
ublisher.

he information herein is offered for informational purposes
olely, and is universal as so. The presentation of the
nformation is without contract or any type of guarantee
ssurance.

he trademarks that are used are without any consent, and the
ublication of the trademark is without permission or backing

by the trademark owner. All trademarks and brands within this book are for clarifying purposes only and are the owned by the owners themselves, not affiliated with this document.

Introduction:

In this book you will discover the best way to watch unlimited TV shows and movies for FREE without any paid subscriptions,

For this you need to have a KODI installed in your device or your computer.

If you own a Fire Stick thin I will recommend the small book which will guide you how to Install Kodi on your fire stick within few minutes.

You can find this book by going here----
>>https://www.amazon.com/dp/B01LXF7MHI

It is assumed that you have already installed kodi app on your laptop or android device.

Now follow the steps laid out in details:

1. First open up XBMC on your device and go to the right where you will find "System".

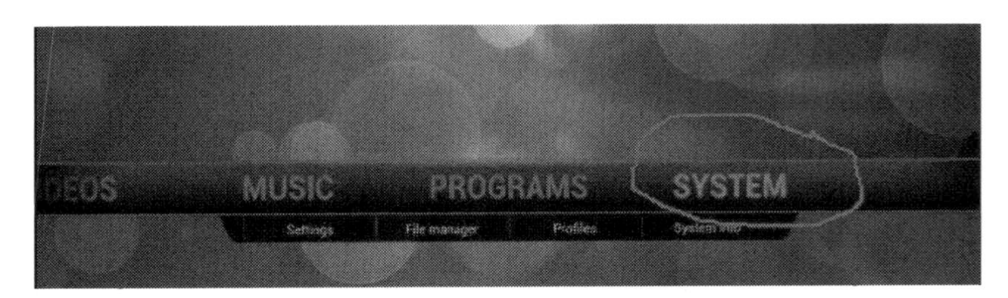

2. Then go to file manager as shown in picture.

. Then go to "Add source" and a window will pop up as shown below and click on "None" as shown below.

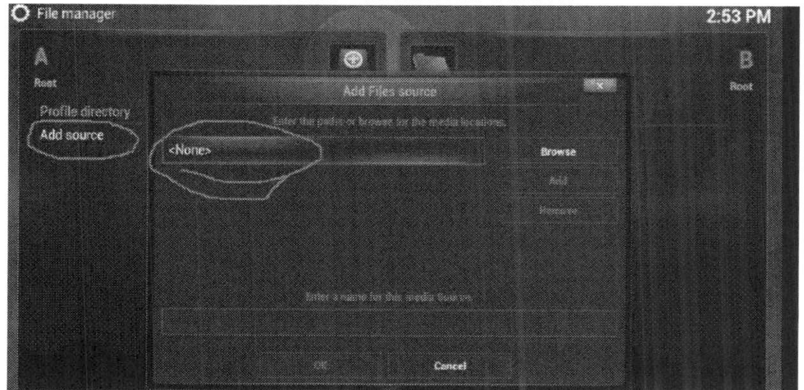

. Then type the following address in the field "http://fusion.tvaddons.ag" make sure to type in the correct address as shown in figure and select "Done"

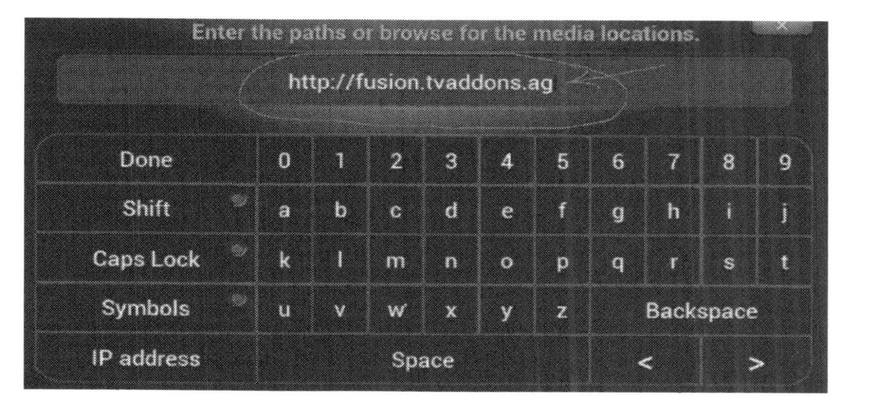

. Now go in bottom as shown in figure

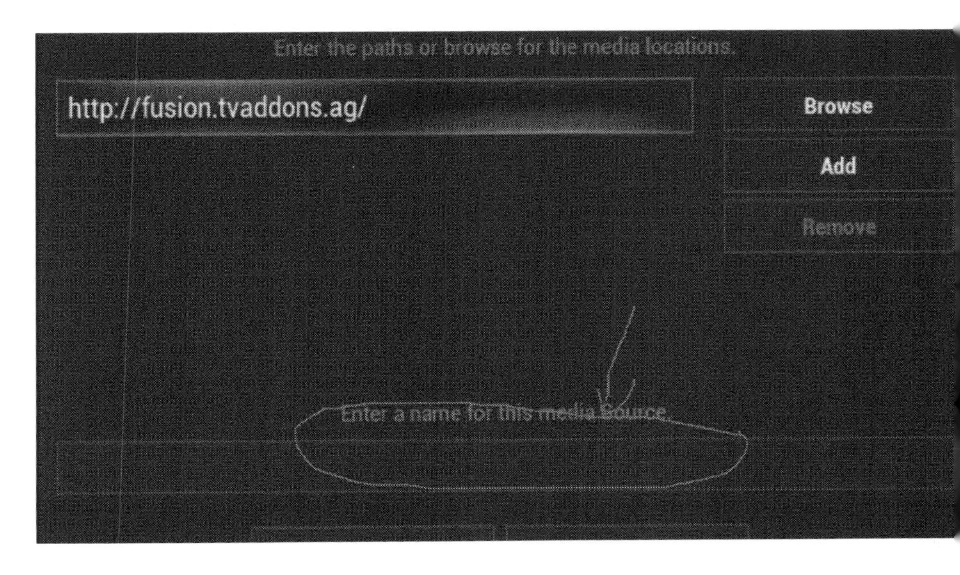

6. Now a window a pops up , give it a name of your choice and click "Done"

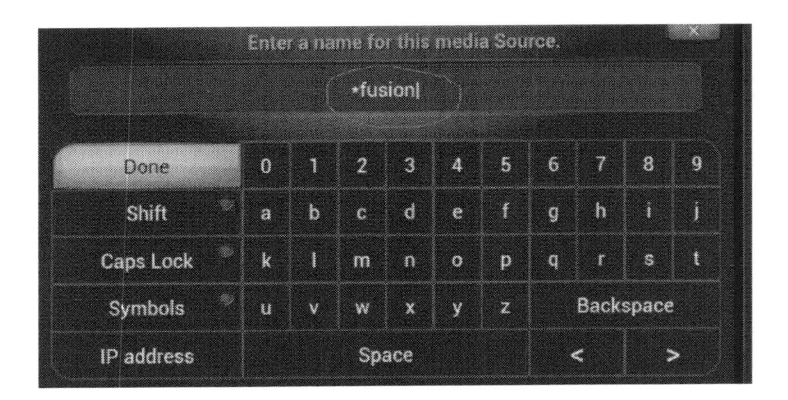

7. Now choose "OK"

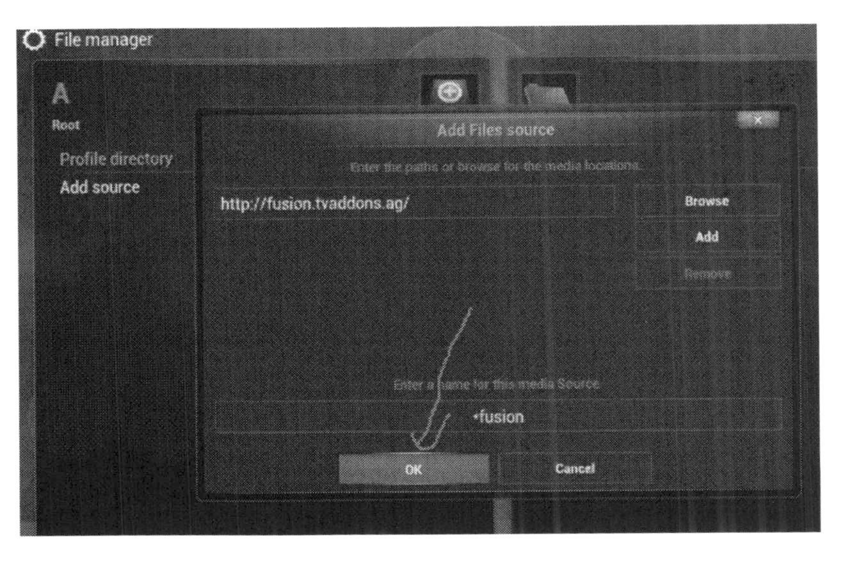

Again go to "Add Source" as shown in picture

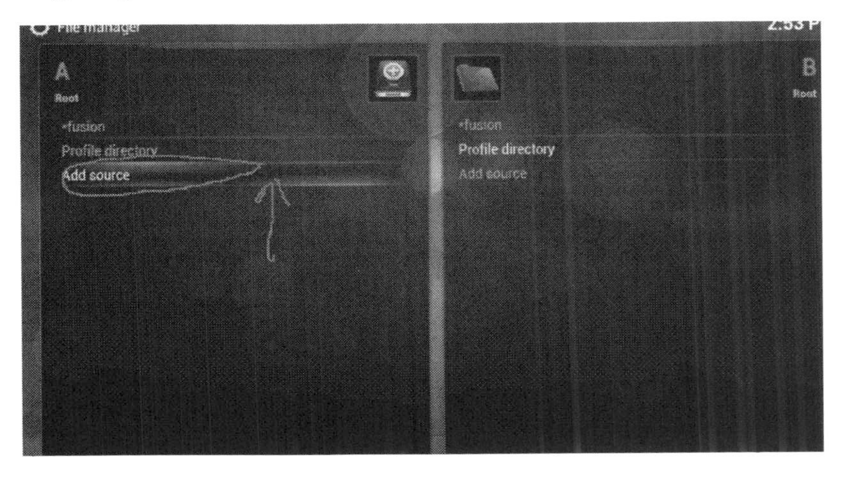

Now Click on "<None>"

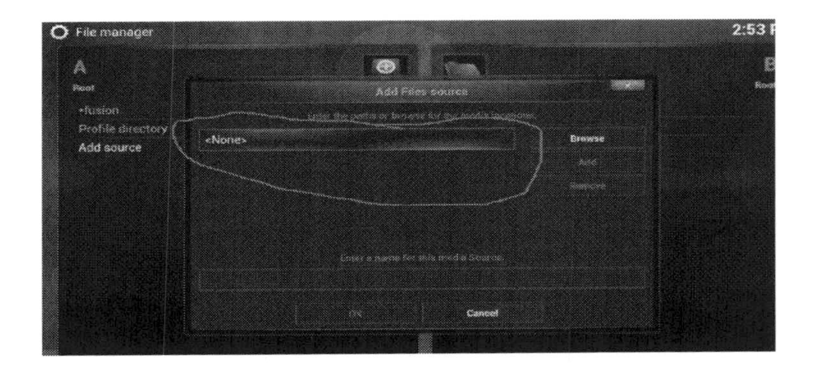

10. Now type in the exact address as shown in figure "http://xfinity.unitytalk.com" and choose "Done"

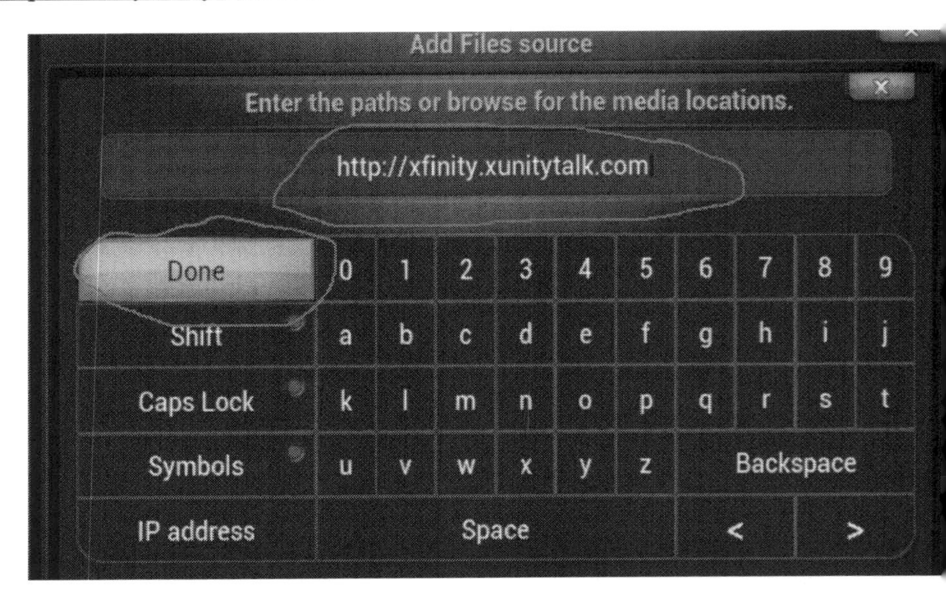

11. Then again go to the bottom of the screen as as shown in figure

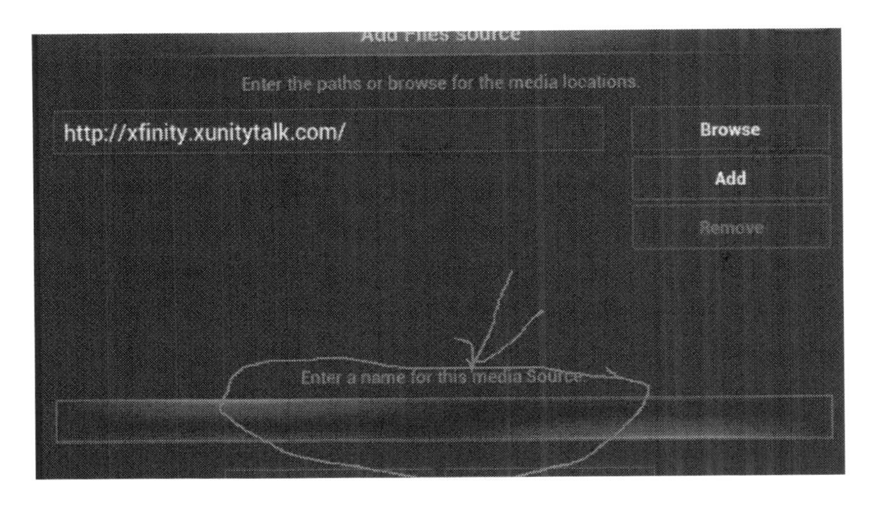

12. give it a name of your choice, it can be anything and hit on "Done" and then choose "OK".

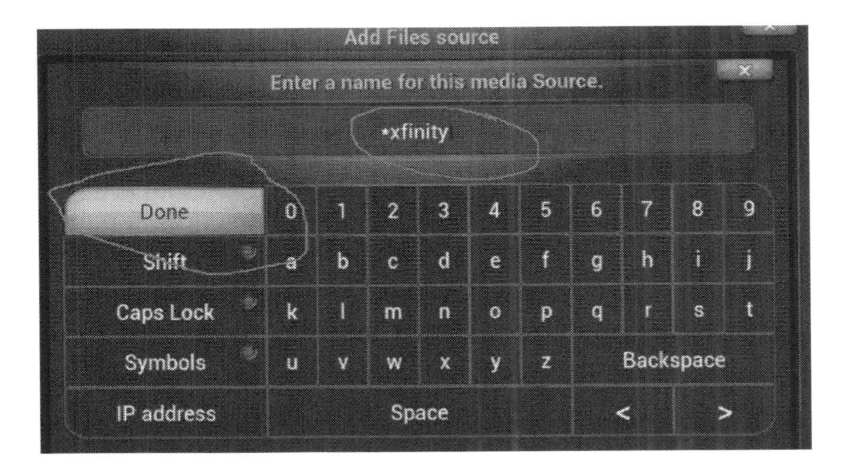

13. Now click back and go to "settings"as shown in figure.

14. Now go to "Add-ons" as circled in figure

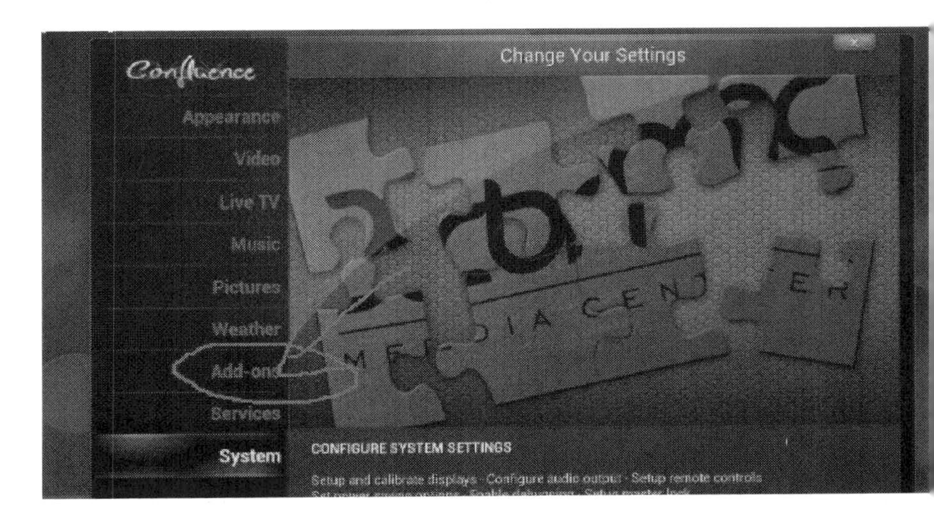

15. Now choose "Install from zip file"

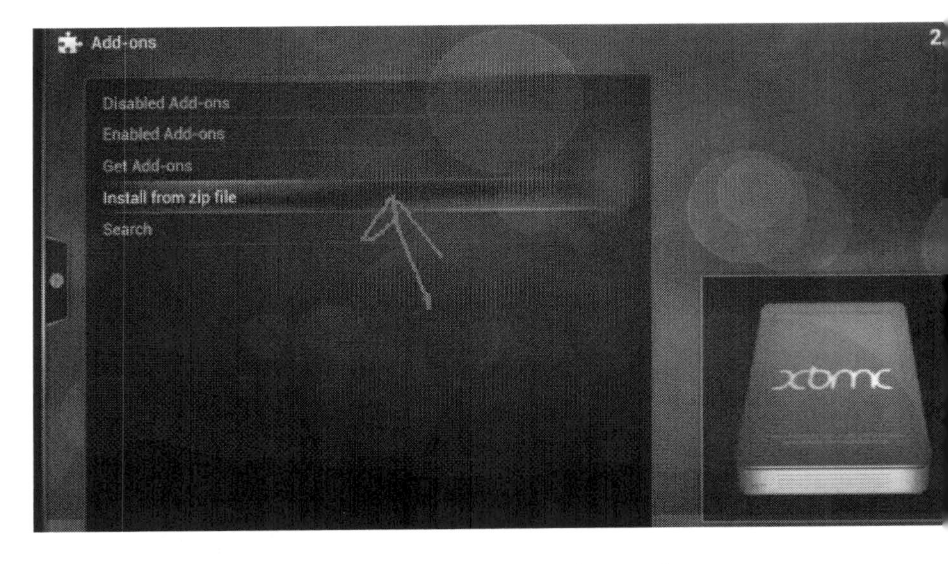

16. Then find the name of your file as you named in in previous step, In my case "*fusion".

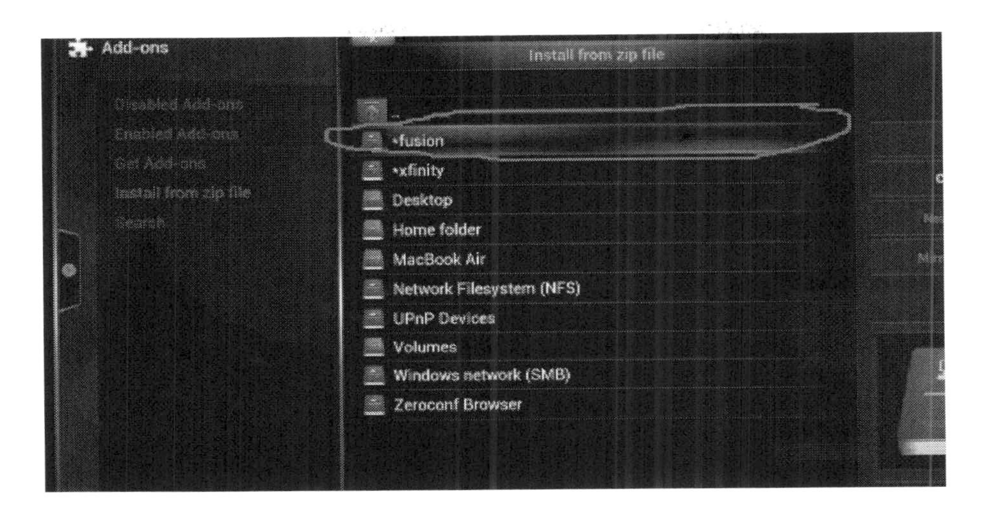

7. Now go to "xbmc-repos" as shown in figure.

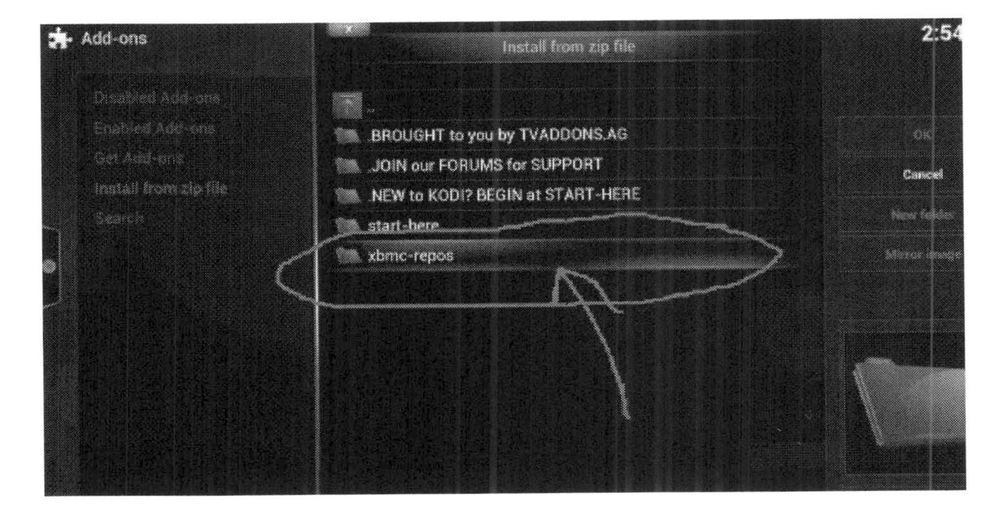

8. Then go to "english"

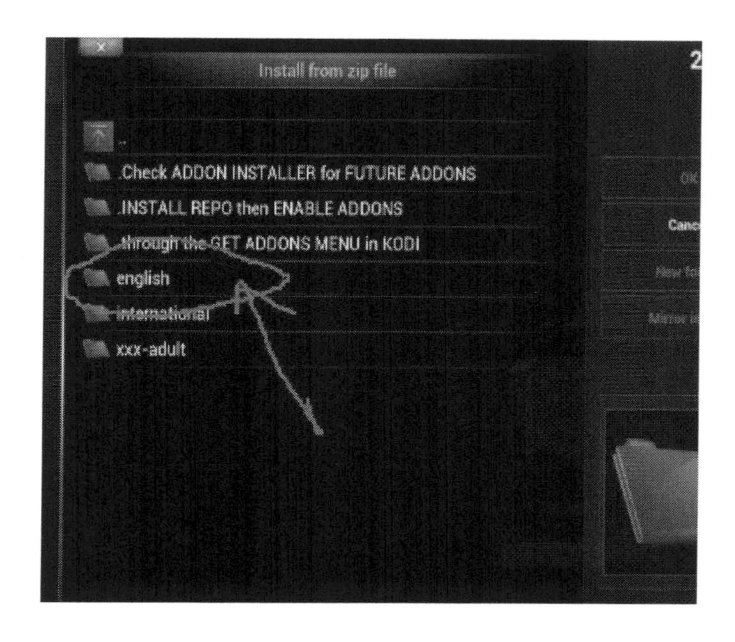

19. go to the bottom and select the file as shown in figure.

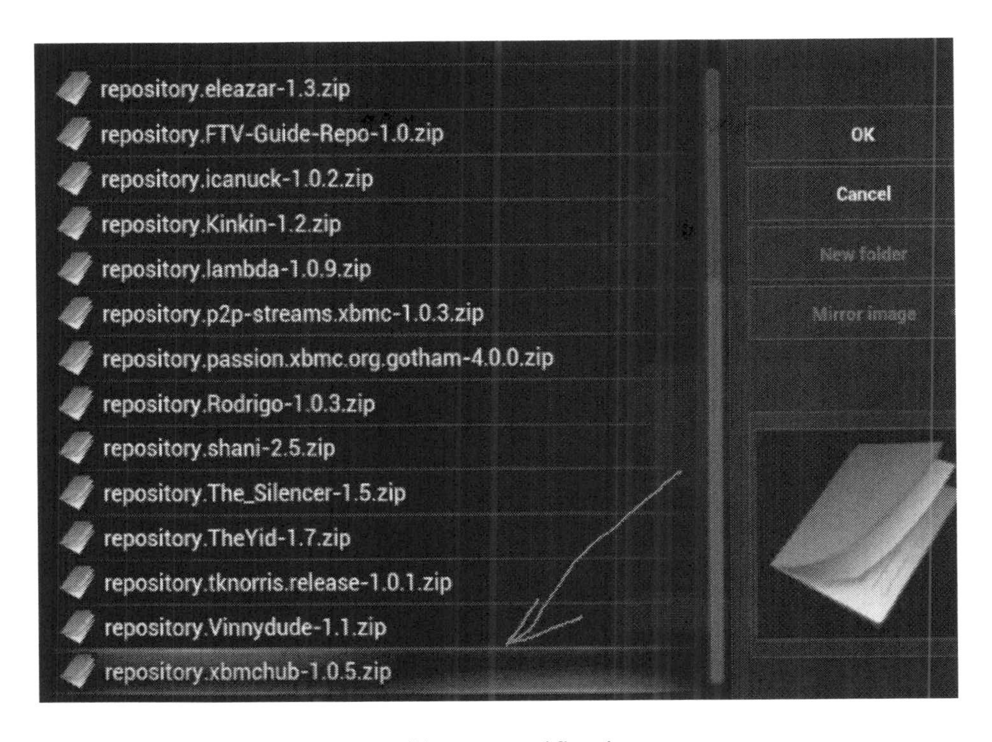

). Now wait for a moment and you will see a notification as circled in figure.

1. Now go back to the previous step and select your other file name you selected as shown in figure.

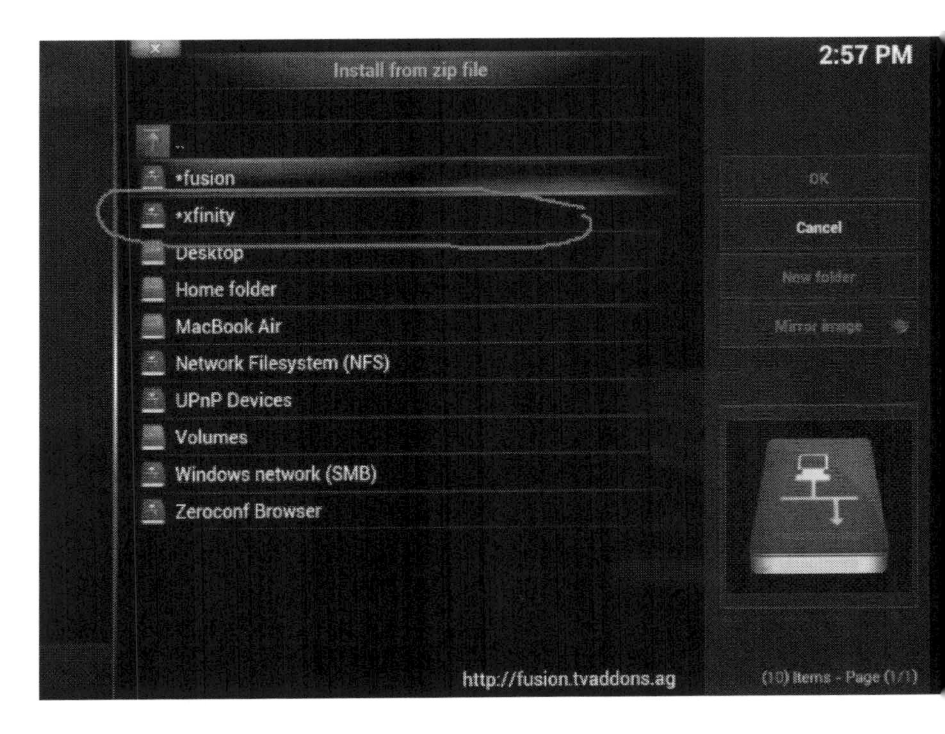

22. Then go to the file "xbmc-repos" a shown in figure

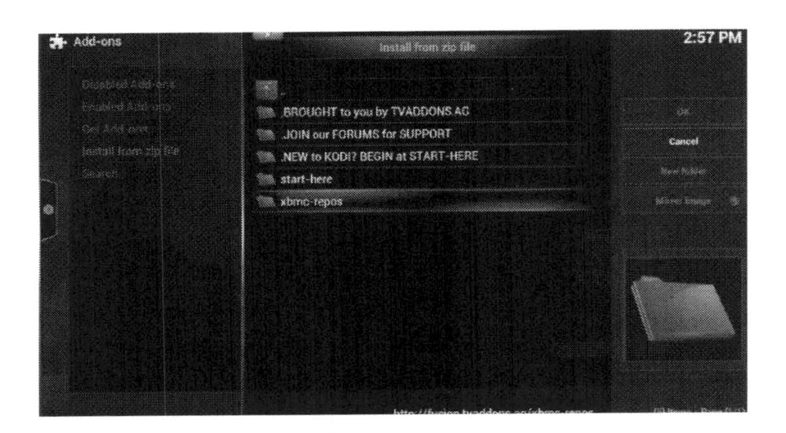

23. Then go to"english"

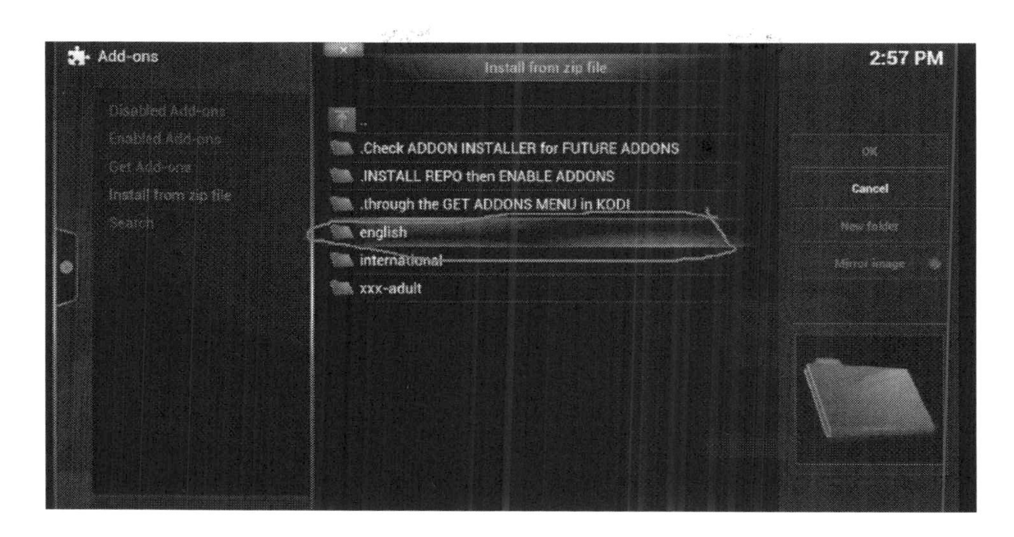

. now go the file as shown in figure

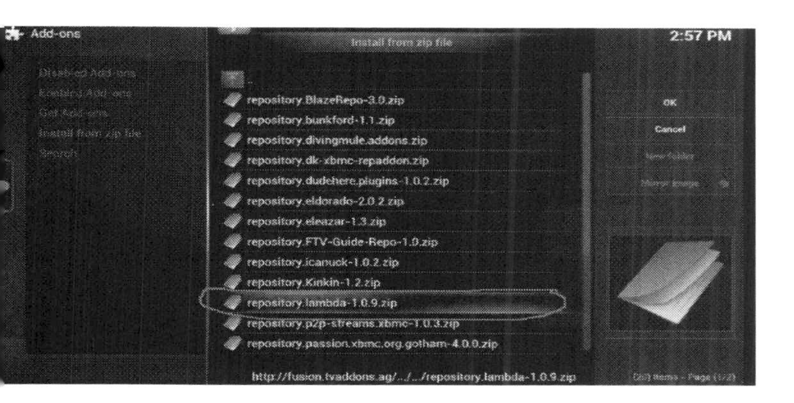

. And wait for the notification to appear on the bottom right a shown below.

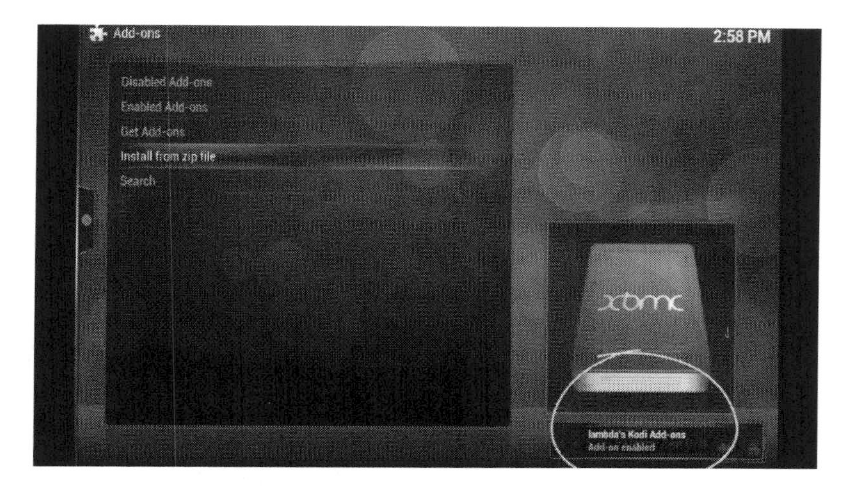

26. Now again "install from zip "

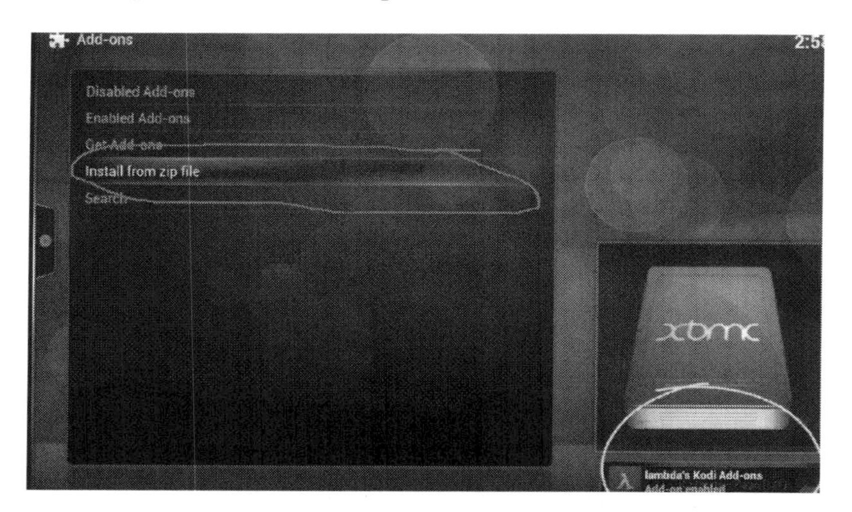

27. and again go to the "*fusion"

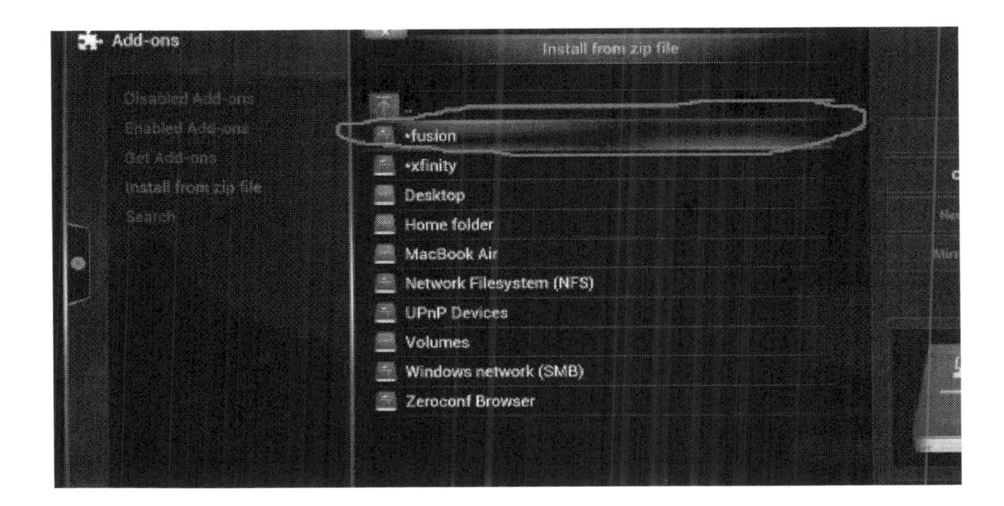

8. go the file as shown in figure.

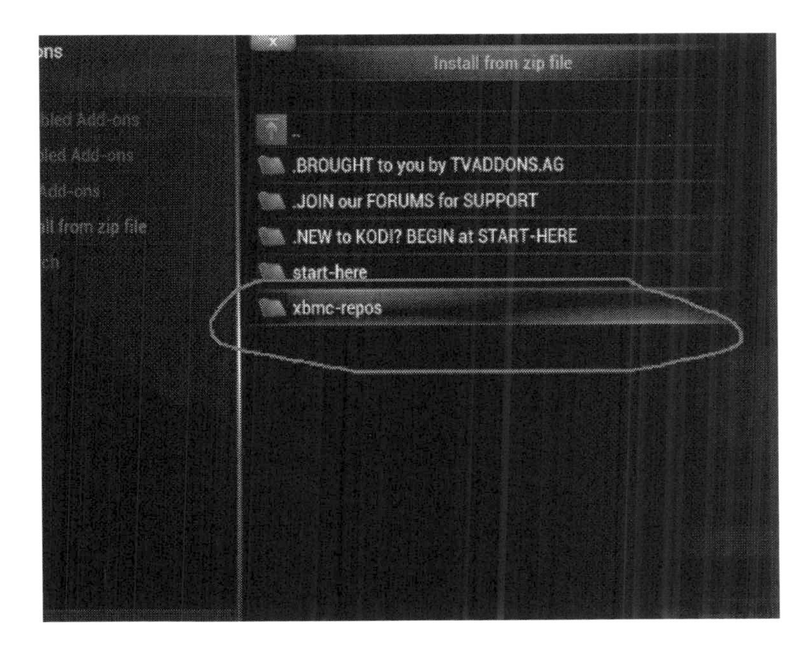

9. And go to "english" again as shown in figure

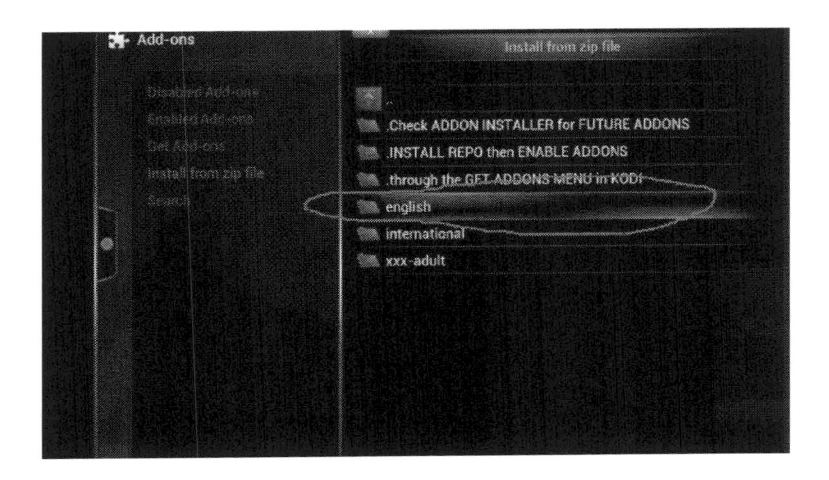

30. Go to the file as shown in figure

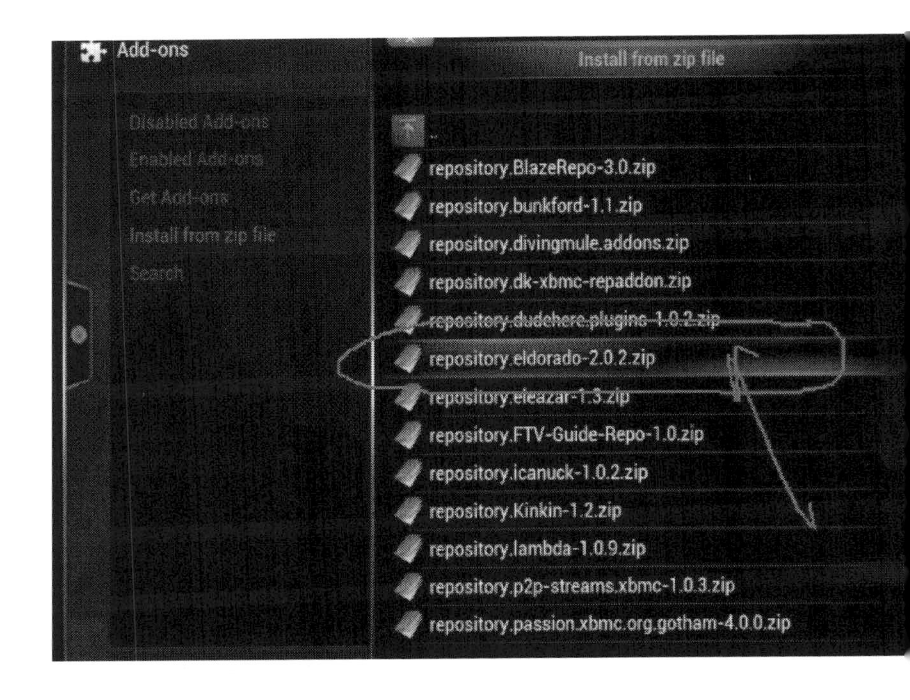

. Now you will see a notification as shown in figure at the bottom right.

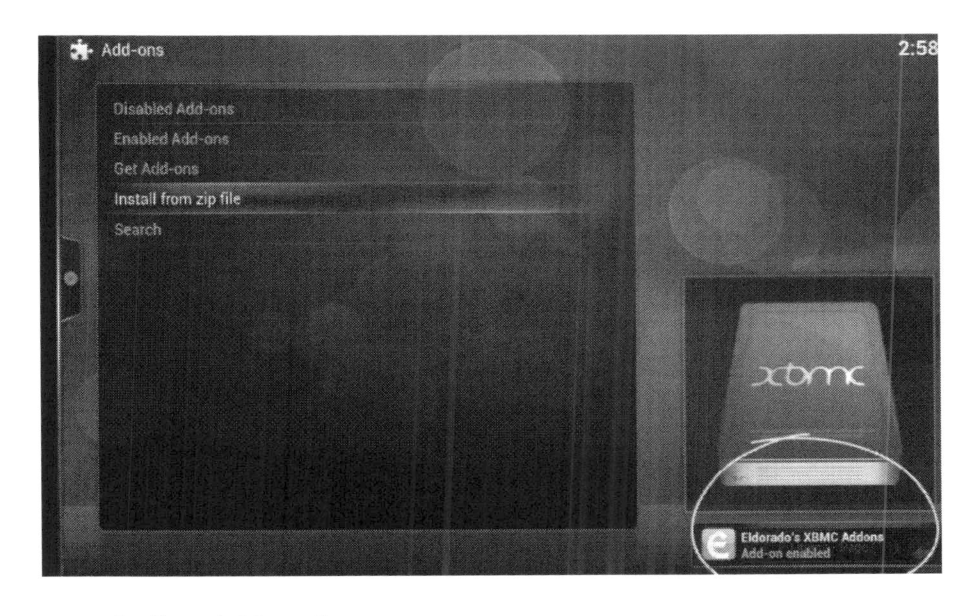

2. Now go to the "get Add-ons"

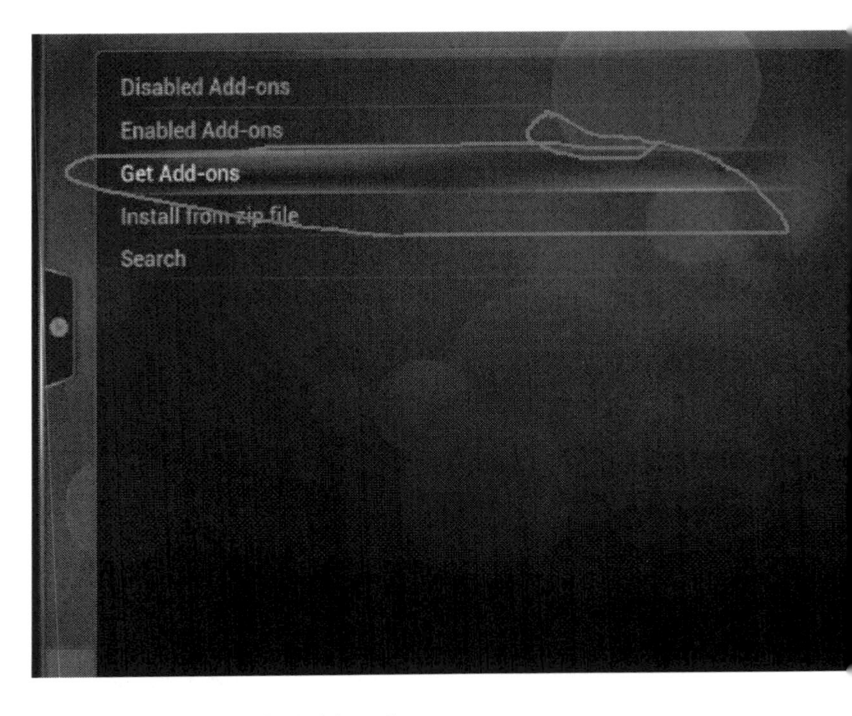

33. Now go to "Eldorad's XMBC Addons"

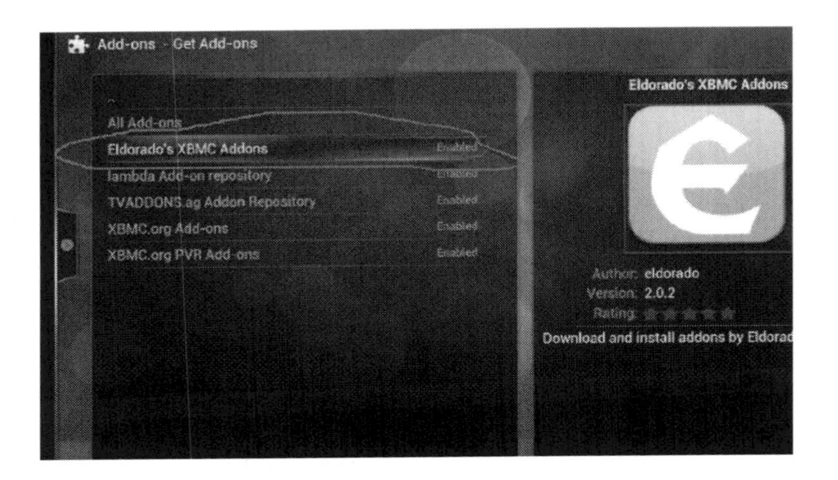

34. Now go to the file name "Video Add-ons"

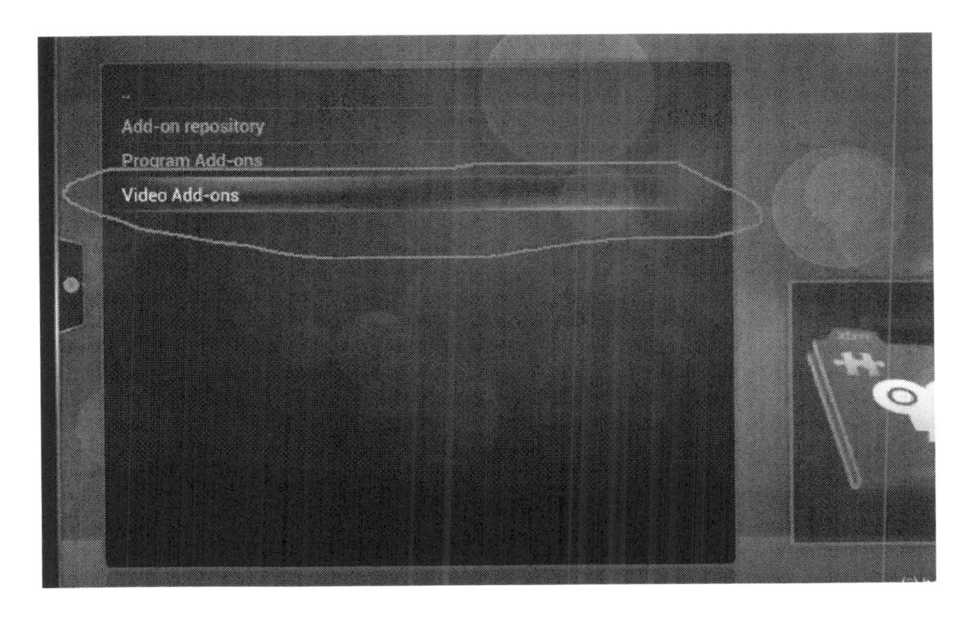

35. Now go to "Icefilms"

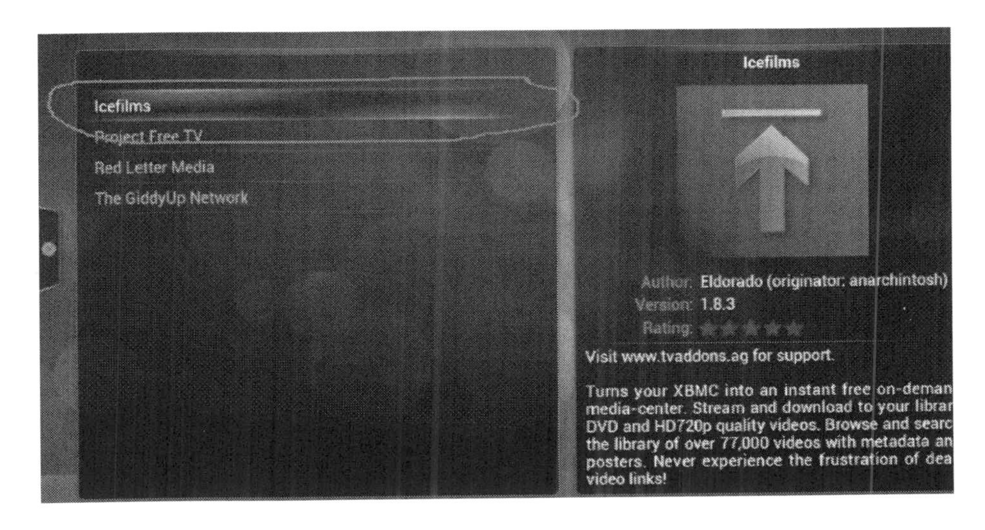

36. and click on "Install"

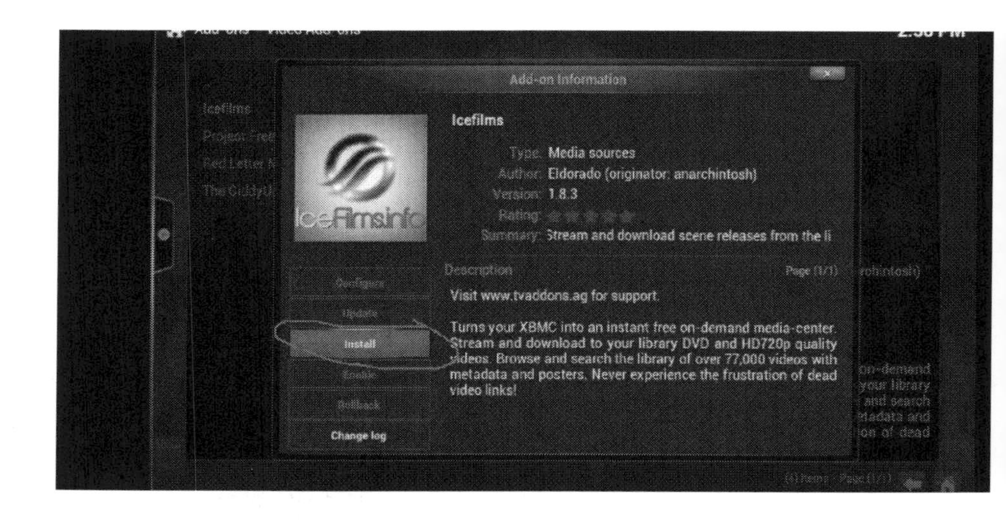

37. you will see the downloading progress as shown in figure.

38. Now go to "The GiddyUp Network"

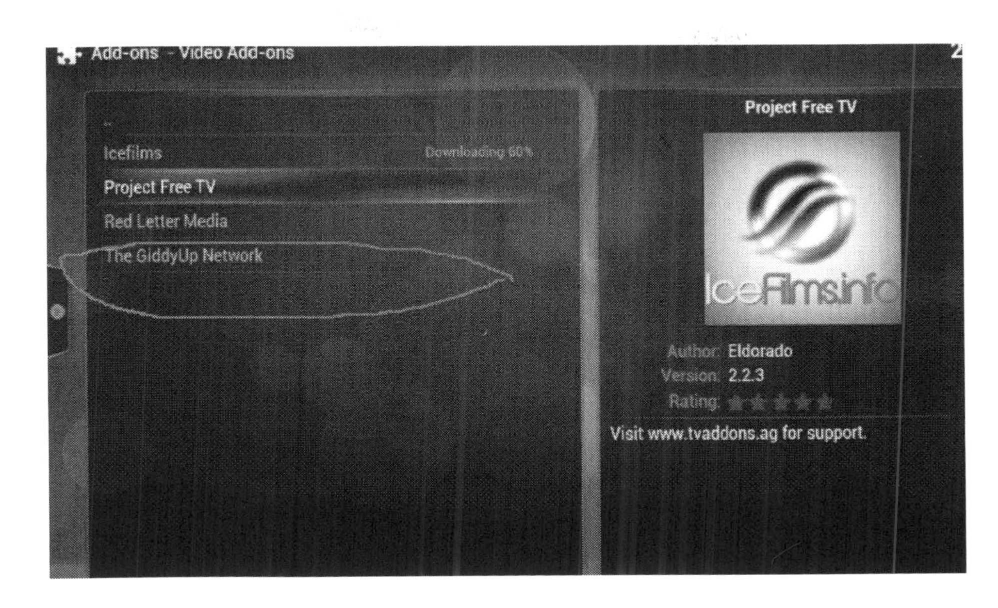

39. Now click on "Install"

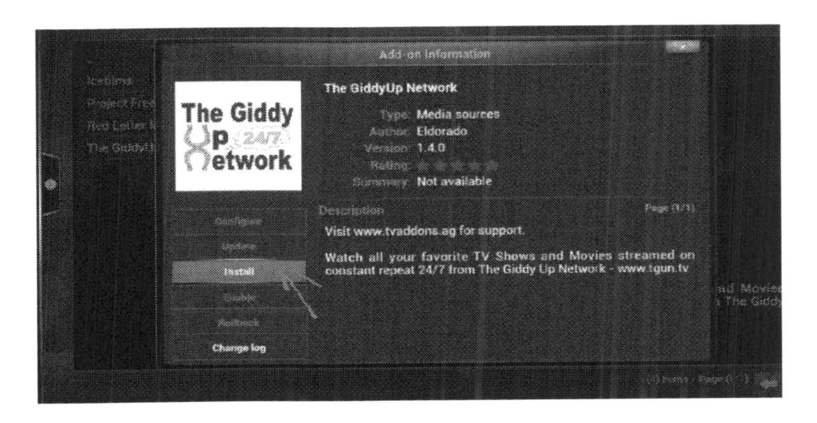

40. Now you will see the "enabled " as shown in figure

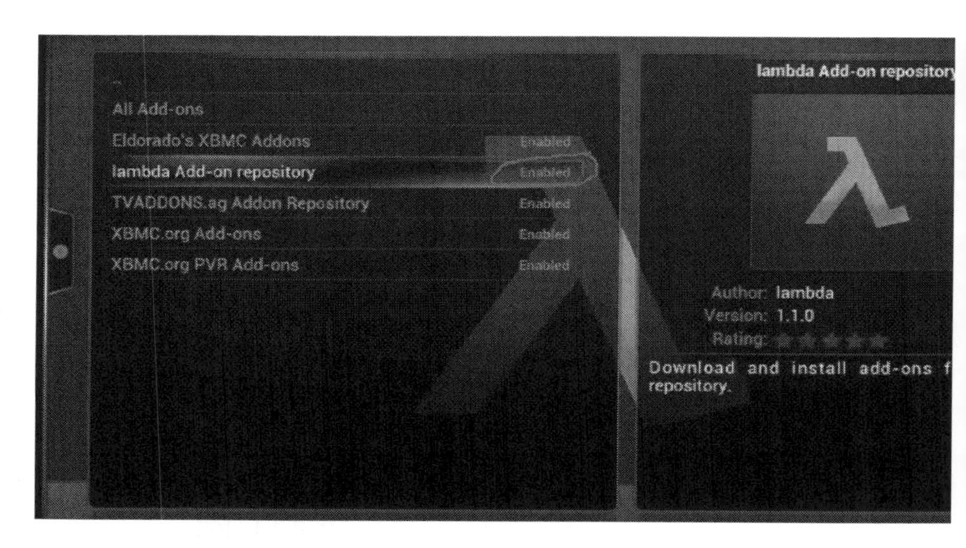

41. Now click back button and go to the directory as you were in step 33. and select "lambda Add-on repos." as shown in figure

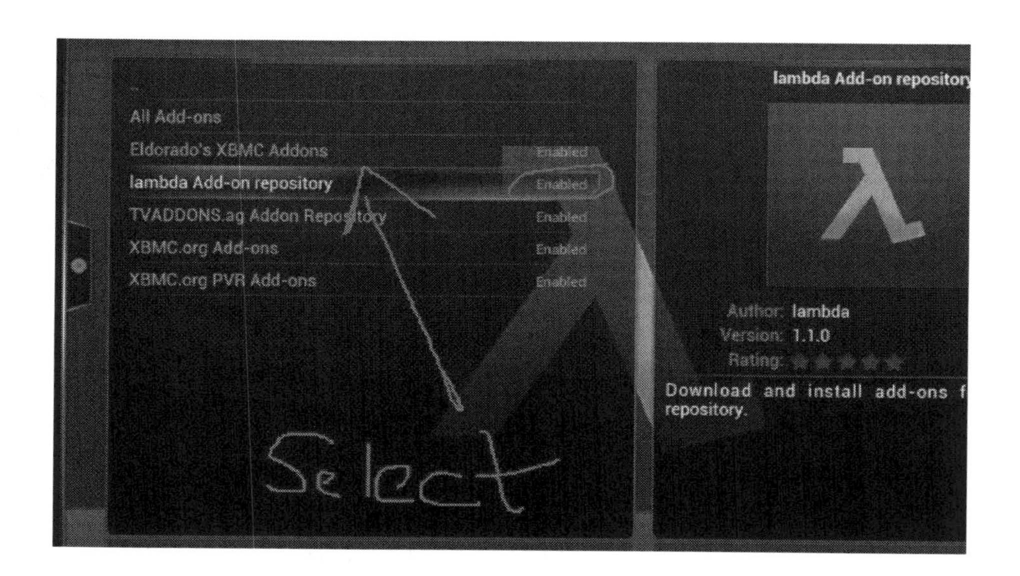

42. now choose "Video Add-ons"

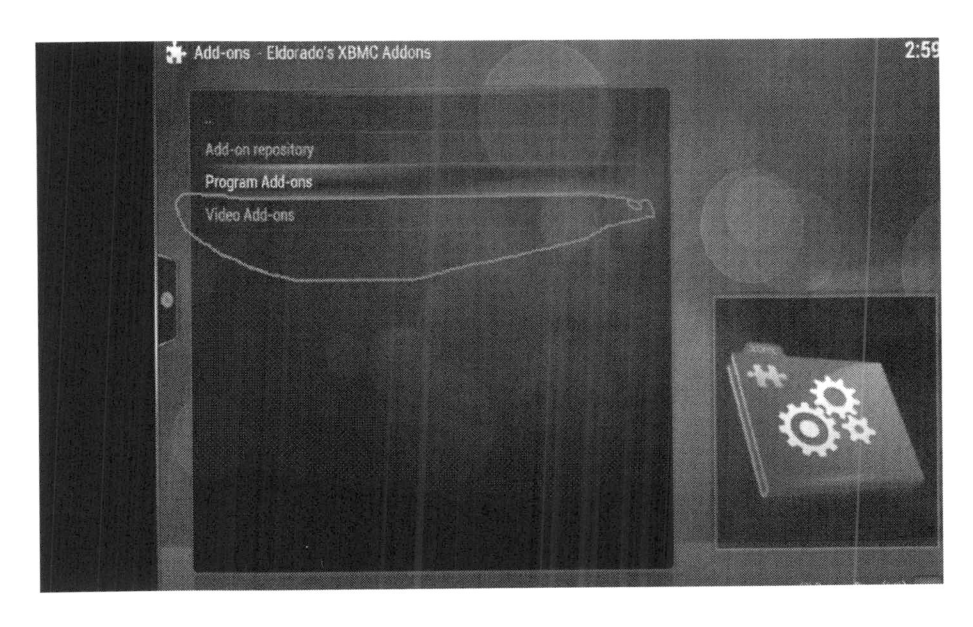

43. now select "Genesis"

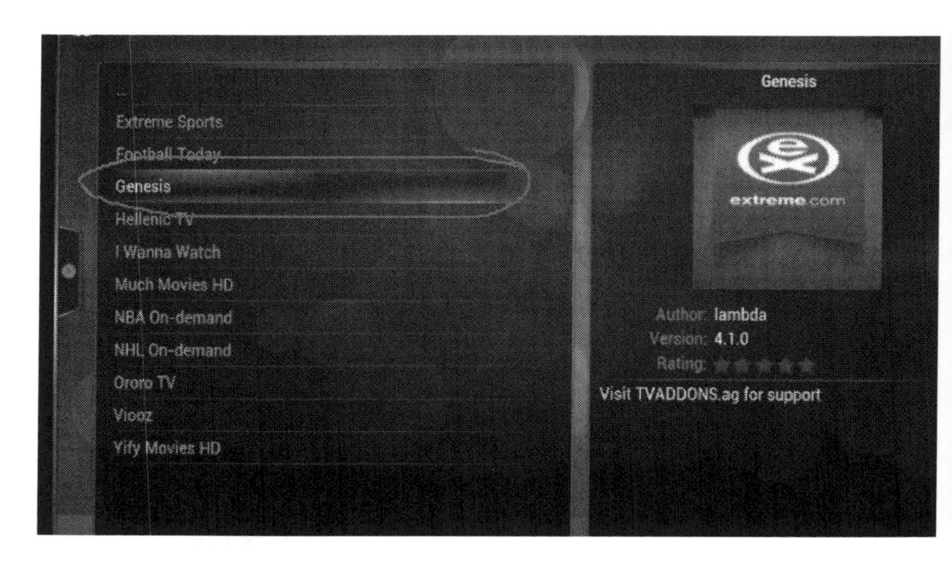

44. and choose "Install"

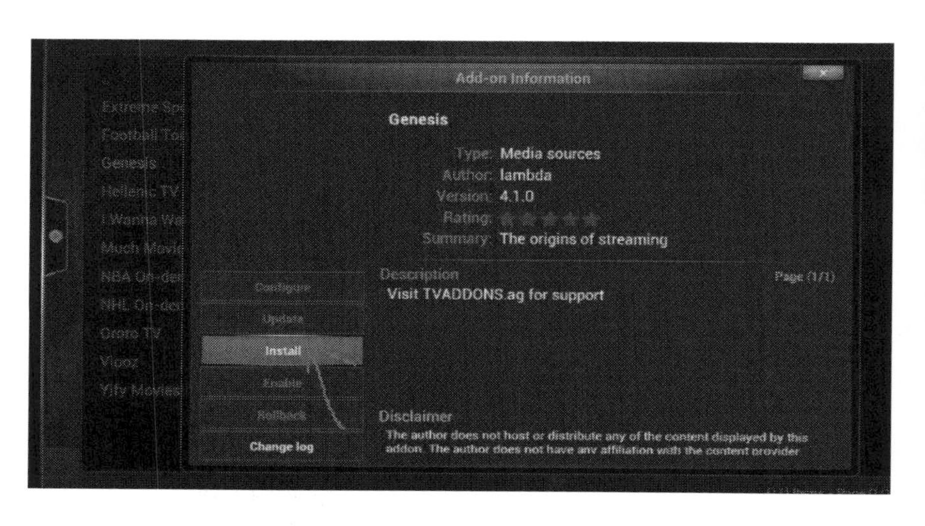

45. now you will see the "Downloading" as shown in figure

46. now select "Ororo TV"

47. and select "Install"

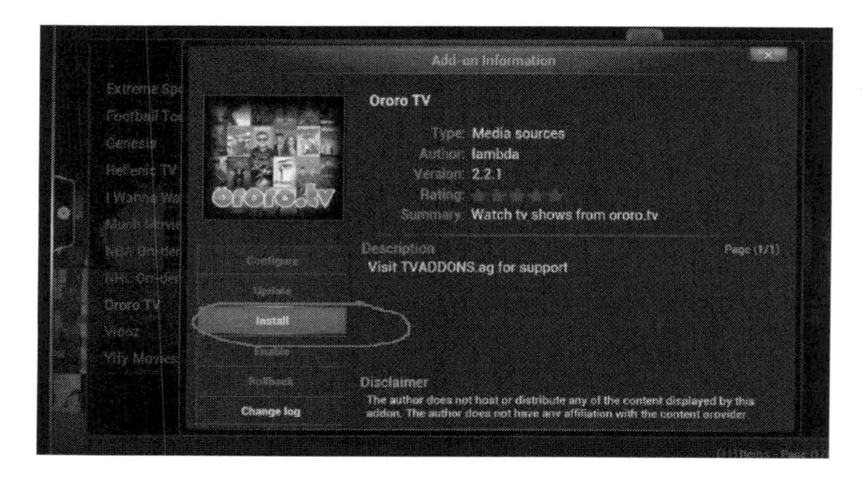

48. Now go back again as you were in step 41 and select file as shown in figure

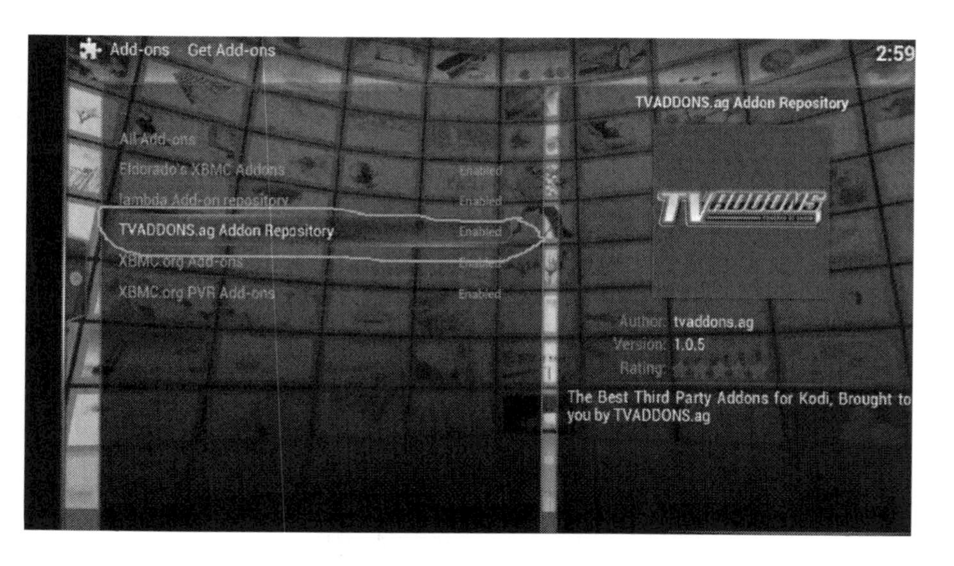

49. now select "Video Add-ons" as below

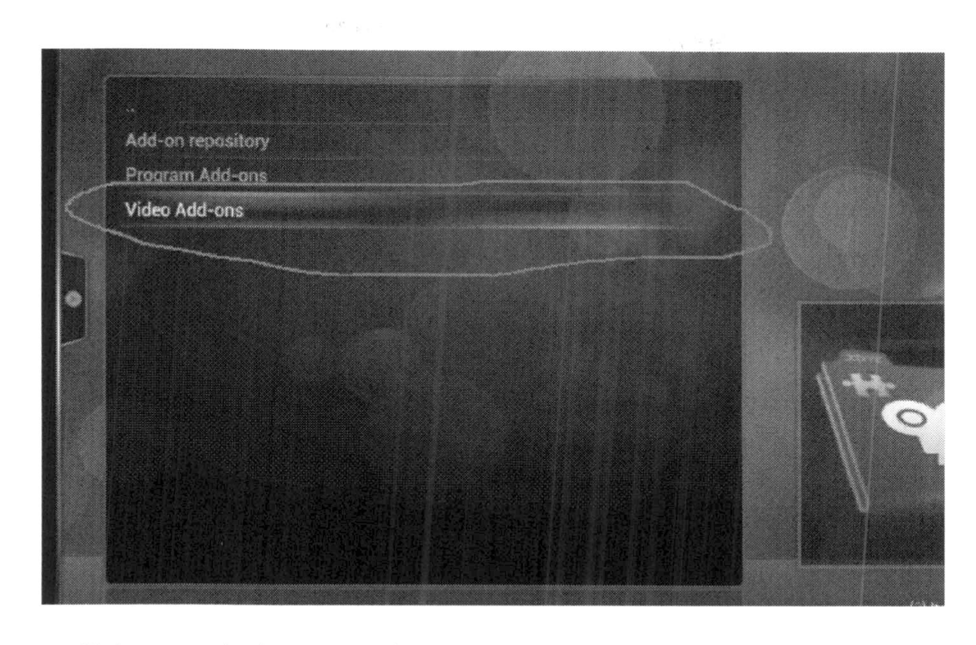

50. now scroll down to the bottom and select "Navi-X"

51. click on install"

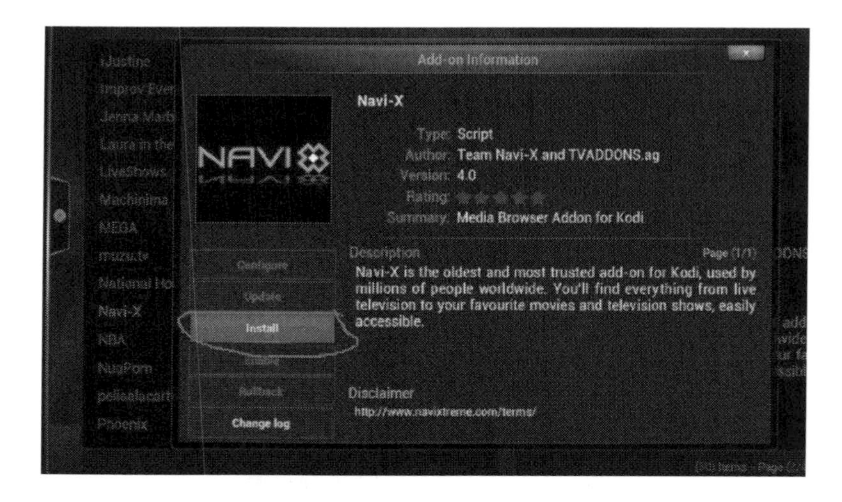

52. Now go back to the system as shown and select settings

53. Now click on "Add-ons"

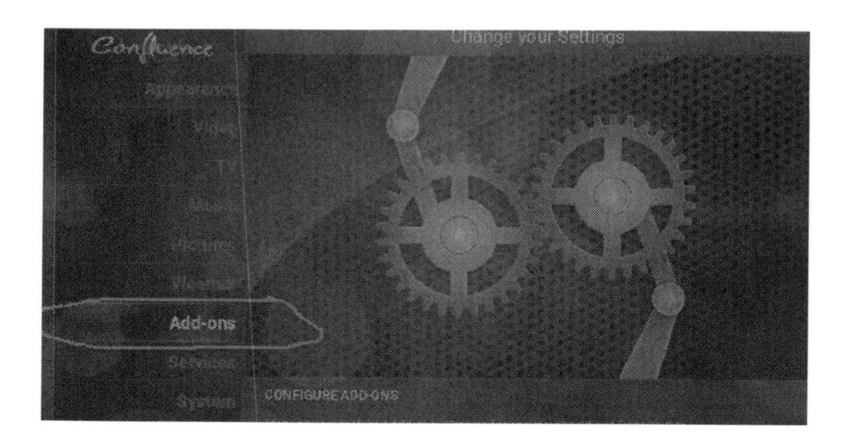

54. Click on "Install from zip file"

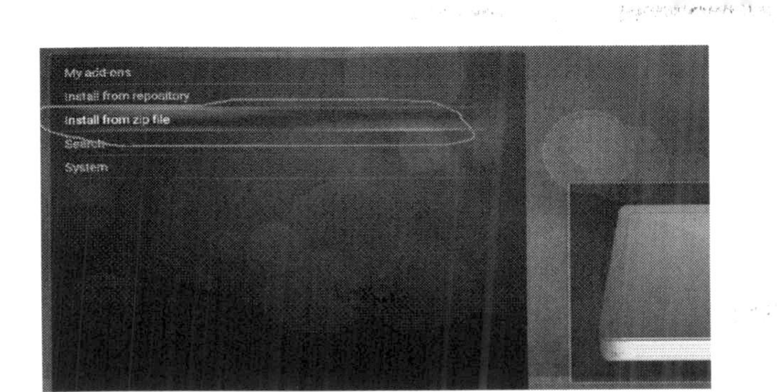

55. Now click the file name "*fusion"

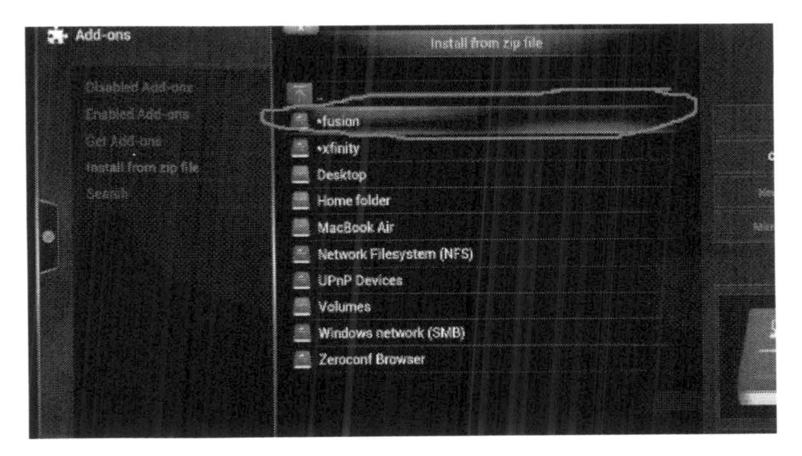

56. click on "xbmc-repos"

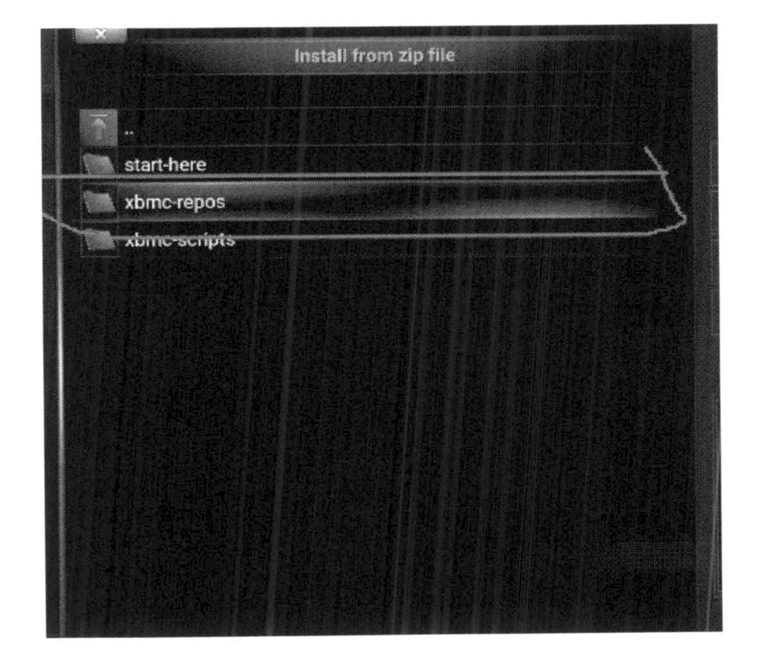

57. now click on "english"

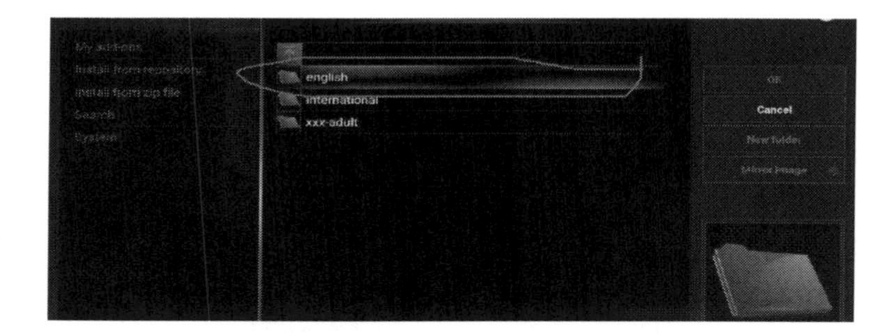

58. scroll to find the file as shown below

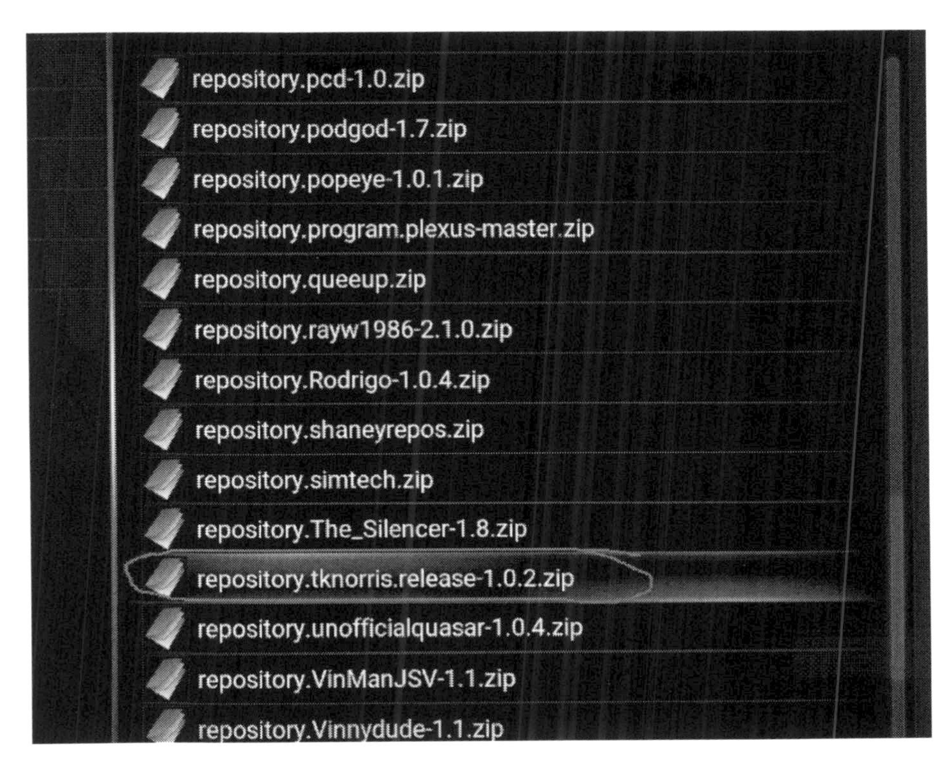

59. now you will see an add on enabled notification at the bottom right.

60. Now go to the screen as shown below and in VIDEOS select Add-ons as shown

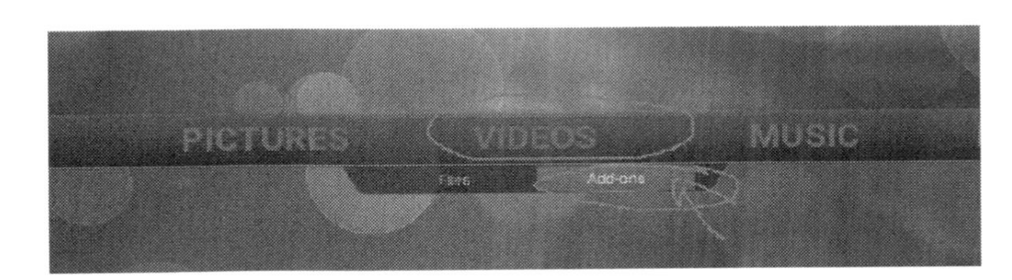

61. click on "get more..."

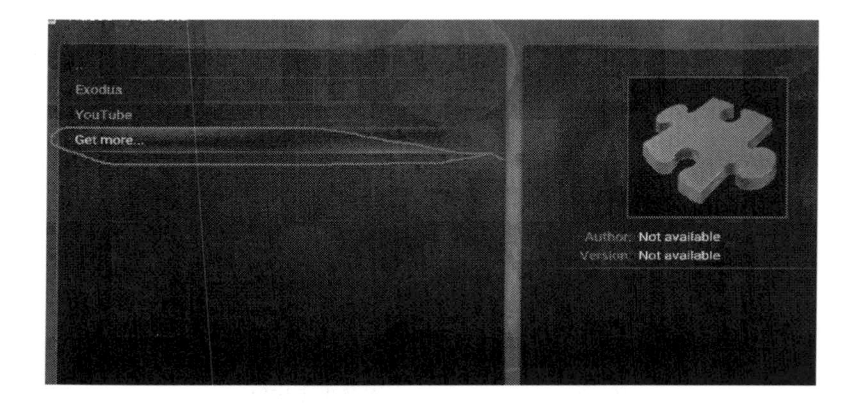

62. Click on 1Channel

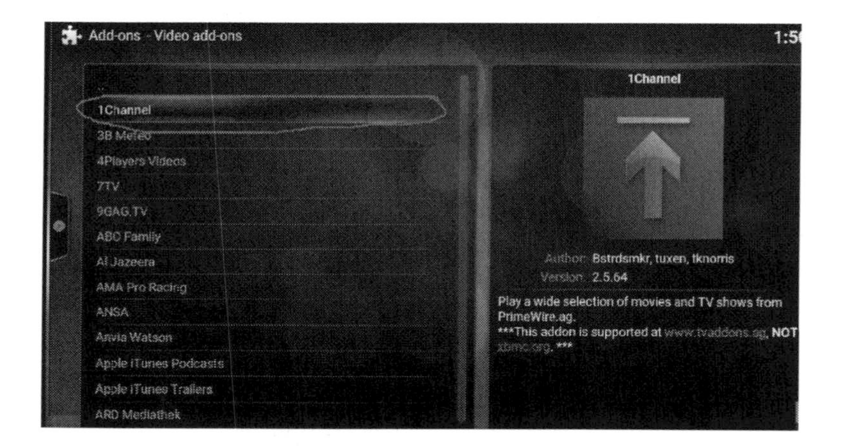

63. click on "Install"

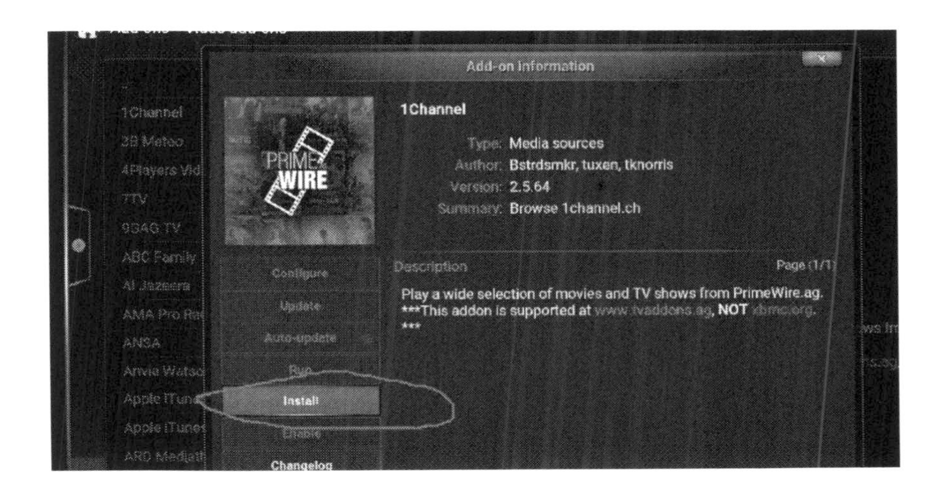

Now that you have installed all necessary Add-ons, now its time to relax and watch your favorite movies and TV shows for FREE.

Simply go to the video Add-ons in the videos section of the main screen and open up your Video Add-ons folder and you will see them listed like below:

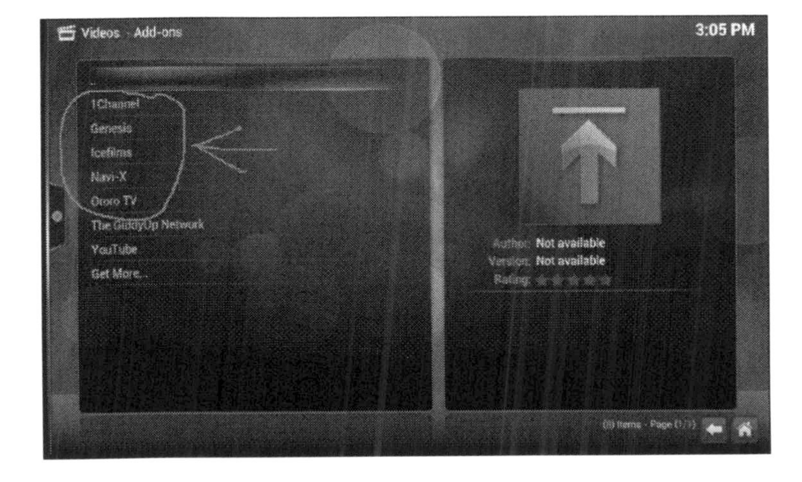

Also make sure to change the view to the "thumbnail" so that it will be viewed like below.

Congratulation you are ready to experience the FREE TV shows and Movies by accessing the Add -ons we downloaded.

Simply go to the Add-ons downloaded one by one and start exploring your favorite movies & Shows.

Conclusion:

If You liked this book then you can leave a review on Amazon so that I can improve the Book
You can leave your feedback here---->>>>
https://www.amazon.com/dp/B01M7O55X8

Happy Watching....

Printed in Great Britain
by Amazon